SOTHIS PRESS

Sothis Press aims to completely reimagine the public domain book with an emphasis on universal access to captivatingly designed print editions. We believe that technology will enable every single book ever published to be perpetually accessible in beautiful print form. Sothis is beauty in ubiquity.

AF811918

www.sothis.press
1 Waterhouse Square, London, ECIN 2ST

Soil Erosion: A National Menace

Hugh Hammond Bennett

SOTHIS PRESS

ISBN 978-1-83548-656-6

This edition is published by
Sothis Press © 2023. The company
is registered in England and Wales.
The Company Number is 14900389.

UNITED STATES
DEPARTMENT OF AGRICULTURE
CIRCULAR No. 33

Washington, D. C. | April, 1928

SOIL EROSION A NATIONAL MENACE

H. H. BENNETT
Soil Scientist, Soil Investigations,
Bureau of Chemistry and Soils

and

W. R. CHAPLINE
Inspector of Grazing, Branch of Research,
Forest Service

UNITED STATES
GOVERNMENT PRINTING OFFICE
WASHINGTON
1928

CIRCULAR No. 33 APRIL, 1928
UNITED STATES DEPARTMENT OF AGRICULTURE
WASHINGTON, D. C.

SOIL EROSION A NATIONAL MENACE

By H. H. Bennett, *Soil Scientist, Soil Investigations, Bureau of Chemistry and Soils*, and W. R. Chapline, *Inspector of Grazing, Branch of Research, Forest Service*

CONTENTS

	Page
Part 1. Some aspects of the wastage caused by soil erosion	1
General statement	2
Some wasting areas	3
Figures on soil wastage	4
Soil erosion not restricted to the South	9
Erosion in the drier regions	11
Relation of sediments to fertility	13
The danger of averages	15
Relation to flood control	16
Limited amount of data available	18
Soil-type information	19
The warning of a geologist	21
Looking forward	22
Part 2. Soil erosion on western grazing lands	24
Introduction	24
Erosion agencies on range lands	25
Natural agencies	25
Part 2. Soil erosion on western grazing lands—Continued.	
Erosion agencies on range lands—Con.	
Fires	25
Mining	26
Overgrazing	26
Other agencies	27
The serious effects of the erosion	28
Flood damage to mountain valleys	28
Silting of reservoirs	29
Reduced productivity of range lands	30
What needs to be done	31
Reestablish the vegetative cover	31
Regulate grazing	32
Protect the cover against fire	33
Aid erosion control by artificial means	33
Conclusion	34
Literature cited	35

PART 1. SOME ASPECTS OF THE WASTAGE CAUSED BY SOIL EROSION[1]

By H. H. Bennett

This circular is concerned chiefly with that part of erosion which exceeds the normal erosion taking place in varying degrees, usually at a slow rate, as the result of artificial disturbance of the vegetative cover and ground equilibrium chiefly through the instrumentality of man and his domestic animals. Removal of forest growth, grass and shrubs and breaking the ground surface by cultivation, the trampling of livestock, etc., accentuate erosion to a degree far beyond that taking place under average natural conditions, especially on those soils that are peculiarly susceptible to rainwash. This speeding up of the washing varies greatly from place to place, according to soil character, climatic conditions, vegetative cover, degree of slope, disturbance of the ground surface, and depletion of the absorptive organic matter in the soil under continuous clean cultivation. Under normal conditions rock decay keeps pace with soil removal in many places; under the artificial conditions referred to, soil removal by the rains exceeds the rate of natural soil formation over a vast area of cultivated lands and grazing lands, often working down to bedrock.

[1] This part discusses only the evils of erosion by rainwash. Much damage is also done by wind erosion, but this phase of the problem is not treated here. The details of checking and preventing erosion and restoring to use the recoverable areas are not included, since that important side of the problem deserves a full paper in itself.

88854°—28——1

GENERAL STATEMENT

Not less than 126,000,000,000 pounds of plant-food material is removed from the fields and pastures of the United States every year. Most of this loss is from cultivated and abandoned fields and overgrazed pastures and ranges. The value of the plant-food elements (considering only phosphorus, potash, and nitrogen) in this waste, as estimated on the basis of the chemical analyses of 389 samples of surface soil, collected throughout the United States, and the recent selling prices of the cheapest forms of fertilizer materials containing these plant nutrients, exceeds $2,000,000,000 annually. Of this amount there is evidence to indicate that at least $200,000,000 can be charged up as a tangible yearly loss to the farmers of the Nation. These calculations do not take into account the losses of lime, magnesia, and sulphur.

In this connection it must be considered that rainwash removes not only the plant-food elements but also the soil itself. The plant-food elements removed by crops (the crops do not take away the soil, but extract nutrients from it) can be restored in the form of fertilizers, manures, and soil-improving crops turned under; but the soil that is washed out of fields can not be restored, except by those exceedingly slow natural processes of soil building that require, in many instances, centuries to develop a comparatively thin layer. It would be entirely impracticable to replace even a small part of the eroded matter, which might be recoverable from stranded material not yet swept into the rivers.

A very considerable part of the wastage of erosion is obviously an immediate loss to the farmer, who in countless instances is in no economic position to stand the loss. Much of the wastage that perhaps might not be classed as an immediate farm loss is nevertheless a loss to posterity, and there are indications that our increasing population may feel acutely the evil effects of this scourge of the land, now largely unrestrained. A considerable part of the erosional débris goes to clog stream channels, to cover fertile alluvium with comparatively infertile sand and other coarse materials assorted from flood water, and to cause productive stream bottoms to become swampy and much less valuable. When the mellow topsoil is gone with its valuable humus and nitrogen, less productive, less permeable, less absorptive, and more intractable material is exposed in its place. As a rule this exposed material is the "raw" subsoil, which must be loosened, aerated, and supplied with the needed humus to put it into the condition best suited to plant growth. This rebuilding of the surface soil requires time, work, and money. In most places, this exposed material is heavier than the original soil, is stiffer, more difficult to plow, less penetrable to plant roots, less absorptive of rainfall and less retentive of that which is absorbed, and apparently its plant-food elements frequently have not been converted into available plant nutrients to anything like the degree that obtains in the displaced surface soil. This comparative inertness of the freshly exposed material is comparable to the lessened productivity brought about in some soils by suddenly plowing large quantities of the subsoil material to the surface. Such raw material must be given more intensive tillage in order to unlock its contained plant food, and on much of it lime and organic manures will be needed in order to reduce its stiffness sufficiently to make it amenable to efficient cultivation, to the establishment of a desirable seed-bed tilth. It bakes easier and, as a consequence, crops growing on it are less resistant to dry seasons, because of rapid evaporation from the hardened surface, and the many cracks that form deep into the subsoil to enlarge the area exposed to direct evaporation. Crops also suffer more in wet seasons because the material becomes more soggy or water-logged than did the original soil. On much of it both fertilizer and lime will be required for satisfactory yields.

Certain piedmont areas whose records are known have, within a period of 30 years, lost their topsoil entirely, 10 inches or more of loam and clay loam having been washed off down to the clay subsoil; and on this clay subsoil, substituted for the departed soil, from 400 to 600 pounds of fertilizer are required to produce as much cotton per acre as formerly was grown with 200 to 250 pounds of fertilizer of no better quality.

While these difficulties of tillage and the lowered productivity are being attended to by the farmer in those fields not yet abandoned, the unprotected fields continue to wash. Unfortunately the farmers in many localities are doing little or nothing to stop the wastage and much to accentuate it. (Pl. 1, A.) In many instances the farmer does not know just what to do to slow down erosion. In many other cases he does not even suspect that the waning productivity of his fields results from any cause other than a natural reduction of the plant-food supply by the crops removed. He does not recognize the fact that gradual erosion, working unceasingly and more or less equally at all points, is the principal thief of the fertility of his soil until spots of subsoil clay or rock begin to appear over the sloping areas.

SOME WASTING AREAS

The southern part of the great Appalachian Valley is an admirable place to see the evil effects of that gradual land washing known as sheet erosion. Here in thousands of areas of formerly rich limestone soil of loam, silt loam, and clay loam texture, the topsoil has been removed. The numerous galls or clay exposures that now splotch the slopes lose their moisture quickly in dry weather. The damaging effects of drought upon crops are felt much quicker than formerly, according to those who have witnessed these changes in the soil. A much lighter rain than formerly now turns the Tennessee River red with wash from the red lands of its drainage basin. Added to the severe impoverishment of a tremendous area of land throughout this great valley, and its extensions southward into Georgia and Alabama and northward into Virginia, are the gullied areas, which are severely impaired or completely ruined by erosional ravines that finger out through numerous hill slopes and even many undulating valley areas. Field after field has been abandoned to brush, and the destruction continues.

Much erosion of the same type has taken place over the smoother uplands of south-central Kentucky; that is, in the rolling parts of the highland rim country; over much of the Piedmont region, and through many parts of the Appalachian Plateau. Land destruction of even worse types is to be seen in the great region of loessial soils that cover the uplands bordering the Mississippi and Missouri Rivers and many of their tributaries, from Baton Rouge, La., northward. Numerous areas, small and large, have been severely impoverished and even ruined in the famous black lands of Texas. Even the drier lands of the West and the comparatively smooth prairies and plains of the North-Central States have not escaped damage. Erosion is wasting the fertility of the soil and even the whole body of the soil in many places where the slope is sufficient for rain water to run downhill. There are some exceptions to this, or rather some partial exceptions, such as the nearly level lands, the loose, deep sandy lands, the highly absorptive gravelly areas, the loose glacial till and morainic deposits in parts of the northern border of the country, the peculiar red lands of the northern Pacific coastal region, and a few others. Although the total area of these more or less erosion-resistant soils is large, the area of those lands which are susceptible to washing and which are being washed in a wasteful way, more disastrously in some places than in others, is very much larger. Save when the fields are frozen or are covered with a blanket of hardened snow, erosion goes on upon these vulnerable lands during every rain that is sufficiently heavy to cause water to run downhill. Even the gentle spring rains cause some erosion, and the surface water flows away from sloping fields muddied red, yellow, or dun, according to the color of the soils of the neighborhood. This color is caused by soil materials started en route to the sea. Most of this material comes from the surface layer, the richest part of the soil.

FIGURES ON SOIL WASTAGE

The estimate of the quantity of plant-food elements annually lost by erosion, as given above, is a minimum estimate based upon a yearly discharge of 500,000,000 tons of suspended material into the sea by rivers,[2] plus twice this amount stranded upon lower slopes and deposited over flood plains, in the channels of streams, and even in the basins of reservoirs, where it is not needed and not wanted. Often this overwash does much more damage than good to the lands affected. It gradually reduces reservoir storage capacity and makes water-power plants dependent more and more upon the flow of the stream rather than upon the impounded water.

It is obvious to all who are familiar with field conditions that the amount of erosional débris in transit to the sea, but temporarily stranded on the way, each year very greatly exceeds twice the amount that actually passes out the mouths of rivers into tidewater. Some soil scientists believe the amount thus annually washed out of the fields and pastures and lodged on the way to the oceans is more than a hundred times greater than that actually entering the sea. The figure used above has been used merely because no satisfactory data upon which to base conclusively accurate estimates are available.

The estimates given do not include the dissolved matter which is annually discharged to the sea, a very considerable part of which obviously comes from erosional products. Furthermore, it is not

[2] Dole and Stabler have estimated that 513,000,000 tons of suspended matter and 270,000,000 tons of dissolved matter are transported to tidewater every year by the streams of the United States (*6, p. 83*).[3] T. C. Chamberlin estimates that 1,000,000,000 or more tons of "richest soil matter" are washed into the oceans from the lands of this country every year (*5*).

[3] Italic figures in parentheses refer to "Literature cited," p. 35.

known how much erosional detritus enters the ocean as drag material swept along the bottoms of streams. This material is exceedingly difficult to measure. The débris thus swept along the bottoms of many streams travels rather after the manner of waves or of sand dunes drifting before the wind. This characteristic of many river beds was brought out before a commissioner appointed by the Supreme Court of the United States in the expert testimony relating to the recent Red River boundary dispute between Texas and Oklahoma. Gilbert (*9, p. 11*) makes the following interesting observations regarding the process:

> Some particles of the bed load slide; many roll; the multitude make short skips or leaps, the process being called saltation. Saltation grades into suspension.
>
> When the conditions are such that the bed load is small, the bed is molded into hills, called dunes, which travel downstream. Their mode of advance is like that of eolian dunes, the current eroding their upstream faces and depositing the eroded material on the downstream faces. With any progressive change of conditions tending to increase the load, the dunes eventually disappear and the débris surface becomes smooth. The smooth phase is in turn succeeded by a second rhythmic phase, in which a system of hills travel upstream. These are called antidunes, and their movement is accomplished by erosion on the downstream face and deposition on the upstream face. Both rhythms of débris movement are initiated by rhythms of water movement.

The amount of plant food in this minimum estimate of soil wastage by erosion (1,500,000,000 tons of solid matter annually) amounts to about 126,000,000,000 pounds, on the basis of the average compositions of the soils of the country as computed from chemical analyses of 389 samples of surface soil collected by the Bureau of Soils (1.55 per cent potash, 0.15 per cent phosphoric acid, 0.10 per cent nitrogen, 1.56 per cent lime, and 0.84 per cent magnesia). This is more than twenty-one times the annual net loss due to crops removed (5,900,000,000 pounds, according to the National Industrial Conference Board) (*16*). The amount of phosphoric acid, nitrogen, and potash alone in this annually removed soil material equals 54,000,000,000 pounds. Not all of this wasted plant food is immediately available, of course; but it comes principally from the soil layer, the main feeding reservoir of plants, and for this and for other reasons it is justifiable, doubtless, to consider the bulk of it as essentially representing lost plant food, without any quibbling about part of it having potential value only.

By catching and measuring the run-off and wash-off from a 3.68 per cent slope at the Missouri Agricultural Experiment Station, on the watershed of the Missouri River, it was found that for an average of six years 41.2 tons of soil material were annually washed from 1 acre of land plowed 4 inches deep, and that 68.73 per cent of the rainfall, the total precipitation amounting to 35.87 inches a year, was held back; that is, 24.65 inches of the 35.87 inches of precipitation were temporarily absorbed as an average for the six-year period. From a grass-covered area of the same slope and soil type less than 0.3 ton of solid matter was removed each year (or a total of 1.7 tons in six years), while 88.45 per cent of the rainfall was retained.

In 24 years this rate of erosion would result in the removal of a 7-inch layer of soil from the area tilled 4 inches deep; but for the removal of the same thickness of soil from the grassed area 3,547 years would be required.

At the Spur substation of the Texas Agricultural Experiment Station,[4] in the subhumid part of west Texas, 40.7 tons per acre of soil material were removed from a 2 per cent slope of fallow land by approximately 27 inches of rainfall. Of this precipitation only 55 per cent was retained (at least temporarily) on cultivated bare land of the same soil and slope without terracing, whereas 84 per cent was retained on an area covered with Buffalo grass.

The erosion station[5] in the piedmont region of North Carolina measured from an uncultivated plot a loss of 24.9 tons of solid matter to the acre each year, when the rainfall was only 35.6 inches, as against a normal of 43.9 inches. On the same slope and soil the erosion from grassland that year amounted to only 0.06 ton to the acre. In other words, the grass held back four hundred and fifteen times as much surface soil as was retained on untilled bare ground. It held back two hundred and fifteen times as much soil as was retained in the cotton plots on the same soil, having the same degree of slope. The uncultivated plot retained 64.5 per cent of the rainfall, the cotton plot 74.4 per cent, and grassland 98.5 per cent.

The agricultural scientists at the Missouri Agricultural Experiment Station have this to say of erosion (*7*):

> Most of the worn-out lands of the world are in their present condition because much of the surface soil has washed away, and not because they have been worn out by cropping. Productive soils can be maintained through centuries of farming if serious erosion is prevented. The soils of Missouri have become gradually less fertile during the last one hundred years due in large measure to the excessive cultivation of rolling lands. Many of the most fertile soils in the rolling prairies and timber lands of this state have been kept in corn until the "clay spots" are evident on nearly every hillside. So much soil has been lost from even the more gently rolling parts of the fields that the yields are far below those obtained by our grandfathers who brought the land into cultivation. The erosion of cultivated fields is taking place at such a rate that it is calling for a decided change in our system of soil management. If we are to maintain our acre-yields at a point where crops can be produced at a profit we must make every reasonable effort to reduce the amount of soil fertility that is carried away during heavy rains.
>
> Approximately three-fourths of the area of Missouri is subject to more or less serious erosion. The map . . . shows where these soils are to be found. It will be seen from this map that erosion is serious on many of the most fertile soils of the state. This is particularly true in the rich rolling prairie regions of central and northwest Missouri, where owing to the fertility of these soils much of the land is kept in corn a large part of the time. It must be remembered that not all the soils . . . erode at the same rate . . . in the Ozark region, they [the soils] are largely covered with timber so that erosion cannot be considered a serious problem.

A single county in the southern part of the piedmont region was found by actual survey (*4*) to contain 90,000 acres of land, largely cultivated at one time, which has been permanently ruined by erosion. The whole area has been dissected by gullies, and bedrock is exposed in thousands of places. Here and there islands and peninsulas of arable land have been left between hideous gullies, but most of these remnants are too small to cultivate. The land has been so devastated that it can not be reclaimed to cultivation until centuries of rock decay have restored the soil. It has some value, however, for

---

[4] Preliminary figures furnished by officials of the Texas Agricultural Experiment Station.

[5] BARTEL, F. O. SECOND PROGRESS REPORT, SOIL-EROSION EXPERIMENTS, EXPERIMENT STATION FARM, RALEIGH, N. C. (A project of the Div. Agr. Engin., Bur. Public Roads, in cooperation with the N. C. Dept. Agr.)

A.—Cotton field showing the early development of destructive gullies in the middles of rows extending up and down the slope
B.—Result of uncontrolled erosion on Greenville fine sandy loam
C.—Deep erosion of the lateral extension type in the loessial region of the lower Mississippi Valley

A.—Six-year-old locust trees which have not only arrested washing on what was once a rather severely eroded area in western Tennessee but are also assisting in a gradual improvement of the soil

B.—A cornfield almost obliterated by a blanket of infertile sand assorted and deposited by erosional waters of one heavy rain

C.—Typical erosion on dark Houston clay of the Alabama-Mississippi prairie region

A.—A field in northeastern Kansas, about 12 miles from Missouri River, that was eroded during a single rainy spell

B.—Highly productive Knox silt loam in northeastern Kansas, about 12 miles from Missouri River, severely damaged by incipient gullies, already 12 or 14 inches deep, which had their beginning in wheel tracks of the grain drill. This is the result of a single rainy period, after grain had been sown, fall of 1927. These will probably grow into deep gullies

A.—Stack of wheat straw dumped in gully in field of Marshall silt loam, which during the fall rains of one year caught and held 430 tons of rich soil material washed from adjacent slopes

B.—An apple orchard in the region of loessial soils, in northeastern Kansas. The trunks of the trees and some of the branches have been buried by soil washed from the adjacent uplands, the surface having been raised about 5 feet growing shortleaf pine and for pasture. The extent of this devastated region unfortunately is yearly growing larger.

Another county in the Atlantic Coastal Plain (*12*) has 70,000 acres of former good farm soil, which, since clearing and cultivation, has been gullied beyond repair. In one place where a schoolhouse stood 40 years ago gullies having a depth of 100 feet or more are now found, and these finger through hundreds of acres of land, whose reclamation would baffle human ingenuity. (Pl. 1.)

The most severely eroded parts of this county are described as follows:

> The Rough gullied land includes areas which, as the result of erosion, are so steep and broken as to be unfit for agriculture. Much of the land classified under this head supports forest. Some areas are available for pasture, but a considerable total area is not even suitable for this use, as there are many deep gullies with steep or perpendicular sides on which no vegetation can find a footing. Providence and Trotman "Caves" . . . are examples of such areas. . . .
>
> In the southwestern part of the county in the Patterson Hills and in another large area . . . southwest of Spring Hill Church, a somewhat different condition is encountered. Here the Rough gullied land consists of narrow-topped ridges with precipitous slopes, covered with ferruginous sandstone fragments. No level land is found here and the slopes are generally too steep even to afford good pasture. . . . One of the largest [caves] within the county has developed in the memory of the present generation, having started with the formation of a small gully from the run-off of a barn. The caves, some of which are about 100 feet in depth and from 200 to 500 feet in width, ramify over large areas. There is little possibility of this gullied land being restored to a condition favorable to cultivation.

In the "brown loam" belt skirting the Mississippi bottoms on the east side, county after county includes 10,000, 20,000, or 30,000 acres of land which have been ruined by erosion. (Pl. 1.) Agriculture has been driven out of a very large part of the upland of several counties in northwestern Mississippi by the gullied condition of the upland. Hundreds of farms in these and many other counties of the region have been abandoned to timber and brush. Unfortunately, the kind of timber that has established itself over much of these dissected areas is largely worthless blackjack oak, simply because pine seed have not been distributed to start valuable pine forests or because black locusts have not been planted.[6] (Pl. 2, A.)

Not only have the uplands been widely and disastrously dissected, but large areas of former good alluvial land have been buried beneath infertile sands washed out of those upland gullies (pl. 2, B) which have cut down through the soil strata into Tertiary deposits beneath. Stream channels have been choked with erosional débris, and overflows have become so common that large tracts of highly

[6] In this connection W. R. Mattoon, of the United States Forest Service, says: "The State of Tennessee through its Division of Forestry has aided several hundred farmers and public organizations, particularly in west Tennessee, in checking gully erosion by the planting of black locust. This work has been done on a gradually increasing scale since its inception, about 1913. Practical methods have been developed of planting one-year-old locust seedlings, spaced about six feet apart each way, over the entire wash or gullied area. Preparatory to planting, the gully banks are plowed off and brush dams built across the channels at strategic points to catch the soil. The black locust produces a heavy surface root system adapted to holding the soil, it is a legume and enriches the soil, it is a vigorous grower and endures thin soils, and it ranks as the second most lasting fence-post timber in this country. Black walnut, yellow poplar, pines, and other trees have also been planted. In addition to checking erosion the land is put to profitable use by growing valuable fence posts and other timber crops and the blue grass that invariably comes in supports limited grazing. A large number of farmers by this method have realized excellent money returns from old gullied lands."

productive soil formerly tilled are now nothing more than swamp land.

The stream bottoms throughout the piedmont region from Virginia southward into east-central Alabama have been impaired by this process of overwash to an even greater extent. Here, probably, considerably more than 50 per cent of the bottom land has been converted into a nonarable swampy waste, entirely as the result of deposition of eroded material. In spite of the terracing that has long been practiced on many farms in the southern piedmont region, wastage of good agricultural uplands has gone on at a distressing rate, because many fields were not terraced and many terraces were not maintained. Thus, unleashed erosional waters, performing in the dual rôle of cutting away the topsoil of the uplands and depositing the less fertile assorted constituents of the eroded matter over the stream bottoms, have brought about an enormous amount of land impairment and destruction.

Some streams formerly navigable have been so choked with sand and mud, purely as a result of erosion, that they have not been plied by boats for a generation or more. E. N. Lowe, of the Mississippi Geological Survey, speaking of soil erosion and flood control in the Yazoo drainage basin, said five years ago (*13*):

> In many of our northern uplands [Mississippi] washing of the soil is progressing so rapidly without let or hindrance over large areas, that some necessary measures must be adopted soon to arrest the process, otherwise vast areas of formerly agricultural land will become hopeless wastes. Large areas in at least a dozen upland counties of north-central Mississippi have already reached such a condition of soil depletion that they are now hardly suitable for any kind of agriculture, and their taxable values are reduced accordingly.
>
> The erosion of these uplands has resulted not only in enormous losses of valuable agricultural soils, but also in concomitant stream-filling throughout those areas. Volumes of silt and sand after every heavy shower are poured into the streams from every furrow, gully and rill that trenches the hillsides, resulting in filling of their channels. The obliteration of their channels causes overflow of the streams after any considerable rain, with deposition of sand over valuable bottom lands, often doing irreparable damage.
>
> For years rapid and destructive filling has affected the Coldwater. Forty years ago boats of large size came up the river to Coldwater to load cotton. Now no kind of a boat can come up Coldwater River, so choked is it with sand bars.
>
> The Tallahatchie was formerly a navigable stream. Even as late as 1900 a small steamer drawing four feet of water plied on the Tallahatchie from Batesville downstream. Now the stream is choked with sand bars, and can be easily waded at almost any place.

In the great cotton-producing section of central Texas, known as the black waxy belt, white spots representing exposures of the basal chalk and marl beds that gave rise to the immensely productive black soil of this region, dot the landscape of the rolling areas. The same thing is to be seen in many parts of the Alabama-Mississippi prairie belt. (Pl. 2, C.) These exposures represent the products of erosion—nonarable land that has been substituted for some of the most productive cotton soil of the world. In one county of this region (*8*) 13.5 per cent of the total area was recently mapped as an eroded phase of the valuable Houston clay soil. It was found that much of this had been too severely washed to allow cultivation, whereas the remaining better parts become highly desiccated in dry seasons, giving lighter and lighter yields as the wearing off of the soil progresses.

SOIL EROSION NOT RESTRICTED TO THE SOUTH

The experts who recently completed the soil survey of Doniphan County, in northeastern Kansas, found that an average of at least 6 inches of soil had been removed from the rich uplands of the county. Nearly all tilled slopes have suffered, some much more severely than others because of variations in the surface relief and in the kind of soil. In one place examined an area of original timber had throughout its extent from 12 to 24 inches of rich soil overlying clay subsoil. This surface layer was so rich in humus, so moist and mellow, that it was possible to dig down through the dark-colored permeable soil with the bare hand, even to the depth of the subsoil. Cultivated soil of the same kind, having the same degree of slope, lying in immediate contact with this forested area had in most places no topsoil at all, as the result of erosion, and in some places even the exposed subsoil clay had been eroded off to a depth of 6 inches or more. Indeed, both soil and subsoil, it was found, had been washed off some areas down to the basal limestone that at one time was 4 feet beneath the surface.

These severe effects of erosion were found as a very common condition in a broad belt over the more rolling lands near the Missouri River; indeed, this condition, or a close approximation of it, was found to be the rule, not the exception, through this more rolling belt, where the virgin soil, the Knox and Marshall silt loams, were among the very richest upland soils in the United States. Apple trees were dying on the eroded hilltops where, seemingly, the soil moisture conditions had been unfavorably upset by the removal of the surface soil layer. In the depressions and on the gentler parts of the slopes and the bench positions some of the rich soil from above had lodged. In these places the apple trees were thriving. A farmer in this section said to the soil specialists:

> We have good apples on the deep soil of the flat places, but we have always had good apples in these places. These places did not need any more soil, they were already deep and rich. We want our soil to stay in the orchards and fields, but it is not staying there. In places 4 feet of soil has been washed off the land. The surface of the ground about our house has been gradually lowered more than a foot. I will show you washed places where not even weeds succeeded this year.

This terrific washing of the land has taken place in the memory of men living in the community. The wasted areas adjacent to the forested land referred to above were cleared about 40 years ago, according to the statements of men in the locality who said they had taken part in the clearing.

Wheat, alfalfa, and sweet-clover fields seeded in the fall of 1927 had been severely damaged by the fall rains. (Pl. 3, A.) In places each depression made by the seed drills had been converted into a small rill way or gully, and the wheel tracks of the seeder in some places had grown into ditches (pl. 3, B), which surely will expand rapidly into formidable gullies that will cause eventual abandonment of the areas affected. In small grain, alfalfa, and sweet-clover fields soil in excess of five tons to the acre was swept from the surface of numerous fields on these splendid soils, the Marshall and Knox silt loams, during a single period of rain last fall. In some fields of steeper slope the loss per acre as a result of this single rainy spell was estimated as amounting to fully 40 tons. Much of this erosional débris passed down from the upland slopes into depressions and stream ways to be carried off in flood waters, although large quantities, as was readily determined, were deposited locally over depressional flats and the flood plains of small and large streams, where none of it was needed. In one place a few miles south of Troy, Kans., where newly planted grainfields had been severely dissected by the fall rains of last year a farmer had left a series of wheat-straw stacks in a depression at the foot of converging slopes. Against the upper side of these large quantities of rich silt had lodged, building up small flood plains, or alluvial fans, 4 feet deep in places. (Pl. 4, A.) One of these stacks had caught 430 tons of this rich soil matter during the single short rainy period referred to, even where vast quantities had been swept by to lower levels after the catchment basin formed by the straw bulwark had been filled. These straw stacks represented the sole attempt to check soil erosion that was observed through several Missouri and Kansas counties bordering on the Missouri River.

Along the outer edge of the Missouri River bottoms, in the northern part of Doniphan County, a farmer had constructed an 8-foot embankment some distance out in the bottoms, approximately parallel to the foot of the upland, in order to intercept soil material that was being brought out of the hills by small local streams. This erosional material was covering the farmer's rich Missouri River alluvium (Wabash and Sarpy soils), causing a reduction in the yield of corn, and was continually washing over the roadways, rendering them impassable. Within 10 years eroded material from the uplands had lodged here level with the dikes, from 5 to 7 feet deep. Thus had been formed a terrace averaging 6 feet deep over 40 acres; and the intercepted soil was not so productive as the land it had buried. The weight of this erosional placed material amounted to about 480,000 tons. It had accumulated at the approximate rate of 1,200 tons to the acre each year. It should be observed in this connection that not all of the erosional detritus brought out of the uplands had been held by the dike. At first, part of it had escaped downstream in the conveying flood waters. Finally drainage had been blocked in that direction, whereupon the material began to escape in the transporting water around the upstream end of the diked area.

This sort of thing is taking place in varying degrees up and down the Missouri River and its tributaries, and along many other streams of the central West. Recently, it was necessary for the soil surveyors working in this great region to recognize a new soil type in order to classify and map material derived from the regional uplands by erosion and freshly deposited over older stream alluvium.

In an apple orchard near Lookout Mountain in northeastern Kansas the trunks of the trees had been completely buried by overwash of silt from the adjacent uplands, and the level of the ground was among the branches of the trees. (Pl. 4, B.) The owner of this orchard stated that although the apple trees had not seemed to suffer by the filling in, the uplands had suffered very greatly from the gradual erosion that gave rise to the transported soil.

Near this orchard a gully is now advancing at a minimum rate of 150 feet a year, according to local information. This ravine is

60 or 75 feet deep, nearly 300 feet wide in places and almost three-fourths of a mile long. It is destined to destroy all the farm land in this fertile valley, including the apple orchard, and it may, with its deploying prongs, cut through the local hills.

Already in this new agricultural region fields (pl. 5, A) and even farms are beginning to be abandoned in the more rolling belts near the river, and land is being rapidly impoverished many miles back from the river. Indeed, all cultivated slopes are suffering to some extent. Nothing is being done to slow down the wastage, but considerable to accentuate it. In general, no effort is being made to cultivate along the slope contours; corn rows are run straight up and down hills as often as otherwise. It is a common practice in this region to plow furrows down the slopes in the spring, in order to allow water standing temporarily in corn "middles" to flow out. These furrows commonly develop into gullies that soon grow beyond control at anything like reasonable cost.

There are no terraces in this region; the farmers do not even know what they are. Erosion is gathering momentum. As the more absorptive topsoil is washed off down to the less absorptive subsoil, the rate of wastage increases. So, this region, which has already suffered seriously from rainwash, is really just upon the threshold of the most impoverishing kind of erosional wastage, and nothing is being done to conserve these splendid agricultural lands, the capital of the farmers living on them and a vital heritage to posterity.

It is not to be understood from the above that erosion in the north-central part of the United States is restricted to the Missouri River region. The wastage is taking place generally throughout this great region, most violently, of course, on the sloping areas. Soil displacement by this process is slow on the very extensive flat areas of the prairie regions that formerly were covered by a most efficient soil-conserving mat of native grass; but even here there is a much greater gradual removal of the rich surface material than is commonly recognized. (Pl. 5, B.) Since the clearing of the sloping and rolling areas and the destruction of the virgin sod, much costly washing has taken place in Missouri, Iowa, Nebraska, Illinois, Indiana, Ohio, Wisconsin, and other States. Recent soil surveys in southwestern Wisconsin have shown that the problem of erosion is a most serious one in many localities. It was found that slopes, especially on the Clinton and Boone soils, which were originally timbered or covered with brush, have been seriously gullied and damaged by sheet erosion from rain water and melting snow. Gullying was found even on bench lands of the valleys (Bertrand soils), and here as elsewhere the stream bottoms were being covered by overwash. These latter instances are mentioned to show that soil wash is a land menace even in parts of the northern border States.

EROSION IN THE DRIER REGIONS

Under the light rainfall of the western dry regions one might reasonably conclude, in the absence of the facts, that erosion is of negligible importance in comparison with that taking place in the humid regions. From the viewpoint of the extent of erosion, such a conclusion would be entirely contrary to the facts, at least for very large areas. The rivers entering the Mississippi from the west carry very large amounts of suspended matter. Some have ascribed this to the treeless condition of the western region. Doubtless this is a contributing factor; and certainly the vast extent of land used for clean-cultivated crops in the prairies and in the more humid eastern part of the Plains, is a most important factor. The peculiar structural behavior of the soils in the regions west of the prairies and eastern plain border, near the headwaters of the streams flowing eastward, coupled with the frequent dashing character of the rainfall, is also an important contributing factor to the heavily silted condition of these streams.

Much of the soil of the dry regions upon desiccation assumes a fluffy loosened condition or structure, to a depth of several inches. The structure is so loose, that the naturally pulverized surface material, even of heavy clays, can be scooped up freely with the hand. Heavy downpours cause this chafflike material to be swept ahead of the flowing water until the soil particles have become thoroughly saturated, disintegrated, and finally coalesced to form an emulsion which might appropriately be styled "liquefied" soil. The flow of this is at first slow, but it speeds up as the emulsion becomes thinner with the increasing proportion of water from rainfall.

In the cattle country of the southwestern dry region numerous places were observed and studied last year (1927), where the richest soil of the region, the areas of deeply accumulated valley-filling material (such as the Reeves silty clay loam), had been washed out entirely from valleys that formerly afforded excellent grazing. (Pl. 5, C.) This process was seen in all stages of development. The washes have their beginning in those places where the natural vegetative cover and normal ground equilibrium have been seriously disturbed. (Pl. 6, A.) Most of them have their start in cattle trails and the wheel ruts in roads. (Pl. 6, B.) In one place near Fort Davis, Tex., an area of approximately 1,000 acres had been so riddled by gullies, which had their beginning in a prairie dog town, that the ground, although once excellent grassland, was almost bare of vegetation. In another locality a gully 30 feet deep, 200 feet wide, and nearly a mile long was seen in the place of a former main highway. This gully was still growing, more rapidly than ever, acording to the ranchman who owned the land.

It was observed in this general region that many of the highways are protected from lateral erosion by retard and diversion dikes, and by diversion ditches dug along the slopes above the roadbed. This form of protection has been successful in some places, but not in others, the difference being due to soil variation and difference in adjustment of the ditch grade to the slope. On some soils of high vulnerability to rainwash, these ditches, built to protect the roadbed, have grown into erosional gullies, which have extended so far that the road has been undermined, necessitating its relocation.

The gullies that cut to pieces the valuable valley grazing lands of this dry country usually go down as incisions having perpendicular walls. When the gullies are cut to the underlying gravel or soft material, an undermining process begins that causes huge blocks of the upper strata to cave into the trenches. These blocks melt away rapidly with subsequent floods. In this region a peculiar soil property serves to accentuate erosion on some very extensive and valuable soil types. The clay soil, when it becomes dry, cracks and scales off from the sides of erosional trenches in such a way as to cause one gully to cut through to another. Thus, numerous natural bridges and caves are formed, and these help to speed up the invasion and destruction of the land.

A ranchman near Marfa, Tex., 15 years ago found that an important strip of his valley grazing land was in danger of being destroyed by an enlarging arroyo. To avoid disaster, he threw a small dam from one side of the valley out across the wash, and about halfway across the floor of the valley; from the end of this a wing dam was turned down the valley and carried for a long distance parallel to the arroyo. The result has been that the arroyo has nearly filled in, both above and below the dam, and the increased water carried over that part of the valley floor lying between the wing dam and the foot of the upland has caused greater subsoil storage of moisture, and thereby made the grazing value of the area affected 15 times as great as it was, according to the statement of the rancher. In addition, the alluvial soil, thus enriched in subsoil moisture has produced valuable crops of feed without irrigation, which in this region is very costly.

In parts of the western deserts railroad companies have found it necessary to construct numerous retard and diversion embankments along their road fills to prevent lateral erosion. The Southern Pacific Railroad in the desert between Niland, Calif., and the Colorado River, for example, has protected sections of its roadbed from erosion by a system of A-shaped embankments that catch the water on the upper side of the track and divert it or concentrate it to soundly constructed culverts beneath the track. Thus, a continuous line of earthen embankments connected like a rail fence, has been built to ward off the abrasive effects of silt and sand-laden desert flood water, and this line at no small cost, must be kept in repair against the erosion of the rainy seasons.

Recently, as the writer was informed, a brief heavy rainy spell (11 inches in three days) in the southwestern part of the United States caused deposition of a layer of infertile sandy material over a valuable orange grove. The trees quickly began to show signs of serious injury, and it was necessary to do something about it. It is said to have cost in the neighborhood of $100,000 to haul the deeper deposits of this inert material (that varying from 1 to 2 feet in depth) out of the grove, and to rake back the shallower deposits from the base of the trees.

In orchards observed by the writer last year (1927) in the valley east of Santa Paula, Calif., fruit trees had been planted on well-constructed terraces to prevent erosion, and, in addition, diversion ditches had been dug along the upper side to catch and divert injurious erosional débris coming out of the adjacent shale hills. To protect the diversion ditch itself, eucalyptus and tamarisk trees had been planted, not only along the ditch embankment, but along the hill slopes above the ditch.

RELATION OF SEDIMENTS TO FERTILITY

It is commonly believed that the products of erosion which do not actually go out to sea are not being wasted. It is believed that frequent deposition of flood alluvium enriches the land, and that floods, therefore, are beneficial, in respect to the productivity of the overflowed alluvial plains. There is some truth in this, of course; but in the main the conception is incorrect, and frequently the good accomplished is greatly overestimated. It has already been pointed out how disastrous overwash of inert sand has been to the alluvial lands of the piedmont region and the "brown loam" belt of the lower Mississippi Valley. This same condition, or an approximation of it, applies also to many other parts of the country. In the Ozark region, for example, the bottom lands of many farms, on which there was but little arable soil in the beginning, aside from the bottoms, have been seriously impaired or ruined by overwash of chert gravel washed down from the regional hillsides. The November flood of 1927 in the New England States laid down upon many of the productive bottom lands a blanket of relatively infertile loose sand and gravel, burying meadows and fields. In other parts of the bottoms the soil was ripped out and washed away by the swift, deep flood water.

The beneficial effects of the sediments deposited by the spring floods of the River Nile are often cited. It is not known precisely what the benefit amounts to in terms of money; but there is no doubt that some measure of soil enrichment does follow the floods of that river. It is obvious, also, that some enrichment of the soil is derived from the finer sediments laid down by the flood waters of the Mississippi. However, some damage is occasioned by the deposits of comparatively inert coarse sand scattered about in the "sand blows," or by patchy deposits that take form locally with every flood spreading over that great delta region. However, the damage and destruction to property and planted crops occasioned by the Mississippi floods quite obviously very greatly exceeds the net benefits accruing from sedimentation. When one thinks of possible benefits to alluvial land derived from deposition of flood-water silt, one should not lose sight of the damage done to upstream farm lands by the removal of the silt into the streams. Also the resultant increase in flood volume due to the additions of solid and dissolved products of erosion is dangerous, and one should not overlook the increased rapidity with which rain water flows off those areas denuded of their more absorptive topsoil. The alluvial soils of the flood plain of the Mississippi and most of its tributaries are naturally so rich that most of them could be cropped probably for many generations, without severe impoverishment of the soil. These alluvial soils are deep, many of them very deep, and exceptionally rich in plant food. The average chemical composition of "buckshot" soil samples taken from Coahoma and Issaquena Counties, Miss., is as follows (*3*): 0.28 per cent phosphoric acid; 0.80 per cent potash; 0.81 per cent lime; 1.31 per cent magnesia.

The phosphorus content of this soil, which is by far the most extensive soil of the lower Mississippi flood plain, is nearly twice that of the average surface soil of the country. It also exceeds the average soil considerably in content of organic matter and nitrogen. Material of this exceptionally good fertility extends to a depth of several feet with but slight change. The condition of fertility is so good that new sediments are not particularly needed, although,

A.—Land in northeastern Kansas formerly cultivated but now used for pasture because of gullying and sheet erosion. The gullies are constantly cutting deeper and fingering out over a wide area
B.—Sheet erosion on almost flat rich black Iowa soil where the first evidence of erosion is the exposure of patches of the clay subsoil
C.—Area in western Texas representative of the destructive effects of erosion on the dry lands of the West

A.—Destructive erosion started in a cattle bedding ground in the arid Southwest. Even the grass roots can not hold the doomed land

B.—Gully 12 feet 8 inches deep in the dry trans-Pecos part of Texas. It started in a wheel rut and will result in the destruction of the entire valley of rich soil

A.—In Orangeburg fine sandy loam the underlying loose sandy material, if exposed, washes away from under the productive easily tilled soil, which then falls into the gullies during rains, together with great chunks of subsoil

B.—The Grenada phase of erosion in the "Brown Loam" belt. The gullying and scooping-out process extends almost equally in all directions and lays waste this valuable loessial soil

A.—Absolute desert conditions brought about by erosion on a once-forested area in northern California. Every vestige of the topsoil and much of the subsoil are gone, and deep gullies have been cut into the soft basal rocks

B.—Intermediate stage of land destruction on Susquehanna soil having a stiff clay subsoil. The devastation has almost reached the stage at which any attempt at reclamation will be unprofitable

C.—Properly terraced field in the piedmont region. Unprotected soil of this type depreciates rapidly because of both sheet erosion and gullying with their supply of lime and fresh organic matter, some temporary increase of productivity necessarily will follow the recession of the water. The degree of this increase can not be estimated with much accuracy with the small amount of available data; but probably its tangible money value does not greatly exceed 75 cents or a dollar to the acre of cultivated land each year for several years following a flood. It must be remembered that the Mississippi delta lands, particularly the predominant "buckshot" soil (Sharkey clay), as well as many of the other alluvial soils of the country, are among the richest soils in the world.

THE DANGER OF AVERAGES

The effect of erosion is extremely variable from place to place, on varying soil and varying slope, with varying vegetative cover and method of land usage. Hence, the average depth of surface denudation that has been commonly computed from river discharges alone, means very little. It implies that the surface everywhere, on steep hillsides and flat prairies, on sand dunes, loam and clay, is being planed down at an equal rate. This is far from the truth. The estimate so often read that erosion is lowering the Mississippi Basin at the insignificant rate of 0.0028 inch annually, is not only too small as an average, but since erosion does not operate over large areas of varying soils according to any plan of averages, such a statement is dangerous both for its inaccuracy and complacency.

A most important thing to know about soil erosion is the rate of cutting away the topsoil, and after that the subsoil of the individual soil types, in those regions of the more vulnerable lands, such as the region of loessial soils, the region of the Susquehanna soils, the Knox, Marshall and related soil regions, the Cincinnati soil region, the Houston clay soil region, and the regions where Orangeburg, Decatur, Cecil, Dekalb, Reeves, Vernon, Putnam, Fairmount and numerous other soils are important.

The studies already made in connection with soil-survey work show that there are many types of erosion, due to many variants, that have to do with the process, chief of which are, (1) soil type, (2) degree of slope, (3) climate, (4) vegetative cover, and (5) method of usage. Some soils can be cropped with a fair degree of safety on slopes having a gradient up to about 20 per cent, such as some of the very porous gravelly soils of the chert ridges in the southern part of the Appalachian Valley. On the other hand, some soils can not be cultivated without steady decline due to erosion, even where the slope does not exceed 1 or 2 per cent. The Knox silt loam, for example, is such a soil. On this soil erosion goes on in all tilled fields where there is any slope whatever.

On some soils greatest erosinal damage is done by gullying; on most soils, however, greater wastage results from that slow type of erosion called sheet erosion. On the Cecil soils of the piedmont region deep, broad V-shaped gullies form and finger out rapidly, whereas on the Orangeburg soils, the sides of the ravines are more nearly perpendicular, and they extend by a process of caving, when the loose sand of the substratum is washed out, so that rapid widening and head-on extension takes place. (Pl. 7, A.) On the Grenada soils of the "brown loam" belt, by reason of a compact subsoil layer peculiar to this group of soils, the washing extends more nearly equally in all directions, and rapidly invades broad areas of fine loessial land wherever the erosion has been neglected in its infancy. (Pl. 7, B.) On gravelly red land in northern California, where smelter fumes have annihilated the forests and destroyed almost every vestige of vegetation, extremely deep, narrow ravines have developed, which make travel over these areas difficult and even dangerous. (Pl. 8, A.) On soils like the Susquehanna, in which impervious heavy clay lies near the surface, the material of the cultivated soil is converted quickly into an approximate liquid condition during rains. This causes the surface substance to flow away rapidly. Following this skinning-off process, the exposed stiff clay is attacked by erosion and gradually cut to pieces by gullies that render the land absolutely unfit for further cultivation. (Pl. 8, B.) Numerous other variations of the manner by which soils erode could be given, but this will not be necessary for the purposes of this circular.

Although there is some erosion on most tilled and bare areas, and probably always will be, wherever water runs downhill, provided the soil is not frozen or protected by hard snow, the damage is greatest in the southern and central parts of the Temperate Zone and in the Tropics. So long as the ground is congealed freezing gives practically complete protection, save on those soils that "heave" badly. Slowly falling rains are everywhere much less destructive as an erosional agent than hard, beating rains. For example, no important effect of surface wash is observable on cultivated slopes of the Fairbanks silt loam, a wind-laid soil, in the Tanana Valley in northern Alaska, where the ground is frozen during eight or nine months and the light precipitation occurs almost entirely as drizzling rain and light showers.

By simple and well-known laws of mechanics the erosive power of flowing water increases enormously with increase of slope, but the destruction accomplished varies greatly with the soil type. Deep sandy soils, as a rule, do not wash severely, especially where the subsoil does not consist of impermeable clay or hardpan. However, some areas of sandy land, such as the Norfolk sand, do wash rather badly, and even gully on those slopes where there is impervious clay at a depth of 4 or 5 feet or less, as is true of areas in east Texas having a stiff subsoil like that of the Susquehanna clay.

RELATION TO FLOOD CONTROL

It is obvious that the erosional débris entering the streams adds to the volume of the water. It is equally obvious that those methods of soil conservation which have been found effective in slowing down or controlling soil erosion, chiefly terracing the land and the growing of trees, grass, and other soil-holding plants, are also methods which will cause more water to be retained in the surface soil and to be stored in the subsoil. Terracing of fields and the growing of trees, grasses, and shrubs on idle lands and areas too steep for cultivation, and upon soils that are highly susceptible to washing, as a combination of practices, will, it is believed, have considerable to do with flood reduction by decreasing the runoff and washoff from many land areas. Soil conservation is somewhat synonymous with moisture conservation. Nothing will hold back all the water, of course, but enormous quantities can be held temporarily or stored for summer-crop use, especially in the subhumid regions. At the same time the rich topsoil can be conserved by these proved implements of soil and water conservation. Soil conservation, therefore, should be an important adjunct of any long-continued system of flood control. To those who have seen the water from heavy rains rushing down unprotected cultivated slopes and bare areas, surcharged with soil matter, and carrying even gravel, cobbles, and bowlders, it is not necessary to argue about the effective contribution widespread use of these soil-conserving methods would make toward flood control as supplementary measures to protection with levees, spillways, and reservoirs.

Suspended material to the amount of 428,715,000 tons annually passes out of the mouth of the Mississippi River alone. This is but a part of the solid material that enters the river and its tributaries since much is left stranded somewhere along the pathway to the sea. In considering the relation of this water-transported erosional material to increased floods, it is necessary to take into account its full significance, along with that of a far greater amount stranded between the source of supply and the streams, in its relation to the increased amount of water flowing off the land areas which have contributed the material. So many tons of silt in the river stand, unmistakably, for so many denuded or partly denuded acres of sloping land somewhere upstream—land enabled by its denuded condition to contribute to the stream at a faster rate more of the rain that falls upon it.

In discussing the relation of forest and other forms of vegetative cover to run-off water and floods, it is frequently contended that, although the methods may have value, the time required for a forest to grow up is too great for this means of assistance to have any important relation to flood problems requiring immediate attention. In this connection the fact should not be lost sight of that the roots of trees and of other plants begin to function as effective agents for holding soil against erosion very shortly after the seedling begins to grow. Greatest efficiency in this respect will come, of course, when the forest or other vegetative cover, as grass, bushes, and chaparal, has made sufficient growth to develop an absorptive, spongy cover of vegetable litter. The immediate effectiveness of grass in holding both soil and water has been conclusively shown by results of the erosional test referred to above. It is said by those familiar with early conditions in the Prairie States that before the extensive cultivation of the land the matted turf of the prairies, in many places, hung like canopies over the banks of streams that carried clear water throughout the year. With the breaking of the land this situation was changed. The streams are more frequently dry in summer and are more heavily laden with silt when the rains come.

The following relates to the effects of rains on sloping areas in Orange County, Calif., following removal of a bush growth by fire:[7]

> During the Orange County Farm Bureau Forestry Tour on November 19th, a remarkable demonstration of the effectiveness of chaparral cover in conserving water by preventing destructive erosion was seen at the Harding reservoir.

[7] Information furnished by C. F. Shaw, University of California, in a letter to the writer. Data obtained from Extension Service Report, December, 1927.

88854°—28——3

In October, 1926, the heavy chaparral cover on this watershed was almost entirely destroyed by fire, leaving the slopes unprotected. During November a heavy rain fell during a 24 hour period. Santiago Creek quickly became a turbulent mass of muddy water, containing over 60% solid matter washed from the burn. The Santa Ana River, which in the past had never had a peak flow of more than 8,000 second-feet during similar rains, showed approximately four times that at the height of the flood.

Harding reservoir was completely filled with rocks, silt and ashes from the burn, and a deposit of a half inch to an inch of this material was left over the entire bed of the Santa Ana River when the flood subsided. Other streams in the vicinity, where watersheds were untouched by the fire, showed scarcely any rise at all, and the water in all of them was clear throughout the storm. After many weeks of shoveling and washing, the capacity of Harding reservoir is less than one-fourth of its original volume.

LIMITED AMOUNT OF DATA AVAILABLE

In this country only a limited amount of information has been acquired concerning the rates of erosion on different soil types, the holding effect of terraces of different build or the possibility of reenforcing them with various stabilizers such, perhaps, as grass, shrubs, or vines, and the rate of alluvial deposition under varying conditions. Only three or four soil types of the many involved have had their susceptibility to erosion measured. It will be observed in reading this circular that little information other than estimates and observations have been given. This is because exceedingly little research work has been done on the subject. It is not known, for example, precisely what type of terrace or what degree of terrace slope is most applicable to the loessial soils of the Marshall, Memphis, and Knox series. It is known that some types of terrace have not given entirely satisfactory results on these peculiar friable soils of such exceedingly high silt content and such low content of clay to bind the silt. Possibly the Mangum terrace, if properly modified and given precisely the right slope, would effectively control erosion on these exceedingly vulnerable soils. Information greatly needed in connection with the problem of erosion should be made available through experimentation and research work as speedily as possible. If a particular type of terrace does not hold in one place and does hold elsewhere, the reason for the failure, as well as for its success, should be determined and the significance of the facts turned over to the farmers of the Nation in forms available for practical use.

As a Nation we are doing very little to abate the evil effects of erosion. Every one who knows anything about it admits the problem is a serious one, but few realize how very devastating is the wholesale operation of erosion. There is necessity for a tremendous national awakening to the need for action in bettering our agricultural practices in this connection, and the need is immediate. Terracing of sloping areas to prevent erosion has been carried on for a long time in the southeastern part of the United States. (Pl. 8, C.) Recently use of this method has extended across the Mississippi River and is being extensively and increasingly employed in Texas, Oklahoma, and Arkansas. The Federal land bank at Houston recently adopted the policy of requiring all vulnerable fields to be terraced before money is loaned on the land. The bank has employed an erosion expert, who, according to press dispatches, not only decides whether or not the property upon which a loan is asked needs terracing, but also goes out and instructs the farmer how to build a terrace if he is unacquainted with the engineering side of this method of conserving soil.

In the region north of Oklahoma and Tennessee the farmers, as a rule, do not know what a terrace is. Most of them have never seen one, and many of them have never heard of this valuable method of soil conservation. In other instances they are not used because the farmers have not been convinced that they are needed, having heard so much about plant food stolen by crops and so little about soil stolen bodily by erosion. Again, terraces are not used because the farmers have not known how to construct them.

Terracing is a very practical method of saving the land against rainwash. (Pl. 9, A and B.) In general the construction of these field embankments is not very costly, and when they are properly built they will pay the cost of construction and maintenance many times over. Although all the details relating to the best methods for terracing some of the peculiarly vulnerable soils are not fully known as yet, it is known that the broad-base, variable-graded ridge terrace, known as the Mangum terrace, properly laid out and built, is a highly efficient instrument for protecting vast areas of land now wasting through the effects of sheet erosion operating increasingly on unprotected slopes. Exceedingly steep slopes, of course, can not be saved by any method of terracing (pl. 9, C), save those expensive methods of building rock walls and huge retaining embankments, such as have been made abundant use of in parts of the Mediterranean Basin and other regions of the world where there is no excess of available farm land and where labor is abundant and cheap. In this country these steeper slopes should be used in accordance with their best adaptation from the economic viewpoint of America, i. e., for forestry and grazing. Probably in this country those erosive sloping lands which range in texture from silt loam to clay and, which have bedrock at a depth of a foot or less beneath the surface, should not be cultivated under any circumstances for generations to come, if ever. There are other soil conditions, also, where the land can not be economically saved by terracing, some even where the mere clearing off of the timber may be followed by wasteful washing. (Pl. 10.) All the details can not be given here.

In addition to terraces soil-saving dams, brush fillings and other obstacles to continuous washing have been successfully employed locally in combating erosion.

There are national associations for the preservation of wild flowers and for the preservation and propagation of wild life but none for the preservation of the soil. Conservation of this most fundamental and important of all resources is seldom seriously considered by any one not directly or indirectly associated with the ownership or management of a farm, and it is too infrequently considered even by the farmers themselves. Erosion is a very big problem. It is doubtful if the farmer can handle it alone.

SOIL-TYPE INFORMATION

The kind of information that is most needed about erosion is that which will apply to definite kinds of land—to soil types that vary from place to place, not only in their crop adaptations and requisite methods of cultivation, but in their resistance to erosion and in the means necessary for checking erosion. Any other method of procedure in studying the problem will be, in no small degree, wasted effort, as methods that may apply to one soil may injure a soil of different character.

As has already been pointed out, terraces must be adjusted carefully, not only to soil type but to slope. If the protection embankment is given too much or too little slope, there is danger of breaks and intensified washing that may exceed that prevailing before the terrace was constructed. In one instance terraces built on Granville sandy loam, in the southern piedmont region, broke with the first heavy rainfall, causing almost complete destruction of the area involved; whereas terraces of the same type made on the Wadesboro clay loam at about the same time and with the same slope withstood the rains that destroyed those on the other soil.

In this connection it is pertinent to refer to a statement of G. E. Martin, of Oklahoma (*14*):

> A half finished job of terracing is likely to result in wasted time, wasted effort, and wasted soil, and tends to bring into disrepute the most satisfactory means, so far determined, of preventing the enormous annual loss of soil fertility which now occurs. This loss constitutes a most serious drain upon the agricultural industry. It is very unlikely that any other industry could suffer such severe losses and survive.
>
> The importance of measuring the slope to determine the proper spacing of terraces can hardly be over emphasized. Too heavy a grade or too much fall along the terrace line, can defeat the moisture conservation objective and may result in hillside ditches instead of terraces.

As an illustration of the important rôle soil character plays in determining the rate of soil erosion, comparisons might be made between the results obtained at the erosion station in subhumid west Texas and those obtained in the humid piedmont of North Carolina. On a 2 per cent slope of the Abilene clay loam at Spur, Tex., 41 tons of soil matter were lost by erosion from 1 acre of land with 27 inches of rainfall; whereas at the North Carolina station only 25 tons of soil matter were removed from 1 acre on a 9 per cent slope with 35.6 inches of rainfall. In other words, although the slope in the latter instance was more than four times steeper than that in the former instance, the eroded material was very much less on the steeper slope.

It is not the purpose of this circular to go into the details of methods for preventing soil erosion, but rather to point to the evils of this process of land wastage and to the need for increased practical information and research work relating to the problem. Instructive bulletins have been published by the United States Department of Agriculture and by the States containing details relating to the best-known methods of checking soil erosion and of filling gullies. Bulletin No. 512, Prevention of the Erosion of Farm Lands by Terracing (*18*), published by the United States Department of Agriculture, is especially instructive in connection with the theory and the practical side of terrace construction; and Farmers' Bulletin No. 1386, Terracing Farm Lands (*19*), is another useful bulletin relating to the subject.

THE WARNING OF A GEOLOGIST

In his address before the Conference of the Governors of the United States, held at the White House in 1908, T. C. Chamberlin, of the University of Chicago, had the following to say in relation to erosion (*5*):

Let us turn at once to the basal factor in the problem, the rainfall, the soil, and the soil-wastage, the special theme of this hour. The rainfall is an inherited asset, the soil is an inherited asset . . . but reckless soil-wastage is a serious error. Soils are the product of the atmosphere and its waters modifying the rock surface. When the atmospheric waters have aided the air in producing soil by rock decay they may pass, on the one hand, into plants or back to the surface soil, and thence to the atmosphere by evaporation, or, on the other hand, they may pass on down to the ground-waters and thence into the streams. The alternative is to rush away as foul erosive floods on the surface, wasting soil and plant food, gullying the surface, choking the ravines, flooding the valleys, silting the pools, filling the reservoirs, sweeping out the dams, barring the streams and clogging the harbors. If it shall be found that all or nearly all the waters should go into the soil and thence into the underdrainage, coming out slowly and steadily by seepage and by springs into the streams, clear and pure, these streams should present nearly ideal conditions for water-food, for power, and for navigation. The solution of the soil problems may therefore be, in large part, the solution of the whole complex of problems of which navigation is the last term.

While soils are formed by the action of the atmosphere and its waters on and in the underlying rock (aided by plants and animals), their surfaces are carried away by wind and wash. At any instant, then, the depth of the soil measures the lag of removal behind production.

We have as yet no accurate measure of the rate of soil production. We merely know that it is *very slow*. It varies obviously with the kind of rock. Some of our soils are derived from material already reduced to a finely pulverized condition. Such are the lowland accumulations from highland wash. Such also is the glacial drift, rockflour rasped from the face of the ledge by the glacial file, and ground up with old soils. On such a base of half-prepared material, soils may be developed with relative rapidity; but even on these, when the slope is considerable, wind, wash and cropping remove the surface much too fast for stable fertility. . . . Without any pretensions to a close estimate, I should be unwilling to name a mean rate of soil-formation greater than one foot in 10,000 years on the basis of observation since the glacial period. I suspect that if we could positively determine the time taken in the formation of the four feet of soil next to the rock over our average domain, where such depth obtains, it would be found above rather than below 40,000 years. Under such an estimate, to preserve a good working depth, surface wastage should not exceed some such rate as one inch in a thousand years. If one chooses to indulge in a more liberal estimate of the soil-forming rate, it will still appear, under any intelligent estimate, that surface wastage is a serious menace to the retention of our soils under present modes of management. Historical evidence enforces this danger. In the Orient there are large tracts almost absolutely bare of soil, on which stand ruins implying former flourishing populations. Other long-tilled lands bear similar testimony. It must be noted that more than loss of fertility is here menaced. It is the loss of the soil-body itself, a loss almost beyond repair. When our soils are gone, we too must go, unless we shall find some way to feed on raw rock or its equivalent. The immense tonnage of soil-material carried out to sea annually by our rivers, even when allowance is made for . . . material derived from the river channels, is an impressive warning, of the danger of negligent practices. Nor is this all; the wash from one acre is often made the wastecover for another acre, or for several. Sometimes one's loss is another's gain, but all too frequently one's loss is another's disaster; and the 1,000,000,000 or more tons of richest soil-matter annually carried into the sea by our rivers is the Nation's loss.

Some of the soluble substances . . . formed at the base of soils are necessary plant food, while some are harmful; but what is more to the point, all are harmful if too concentrated. There is need therefore that enough water pass through the forming soil, and on down to the ground-water and out through the underdrainage, to carry away the excess of these products. An essential part of the best adjustment is thus seen to lie in *a proper apportionment of the amount of water which goes through the soils.* If this be not enough, the plants will suffer from saline excess; if it be too much, the plants may suffer from saline deficiency.

Here, then, lie a series of possible excesses and deficiencies; between them falls the golden mean which gives best results. Thus the problem of soil management is a problem of proper balancings and adjustments, a problem to be solved by science and common-sense forced to their best by the energy and intelligence of the American farmer.

The key to the problem of soil conservation lies in due control of the water which falls on each acre. This water is an asset of great possible value. It should be looked upon as such. It should be computed by every acre-owner as a possible value, saved if turned where it will do good, lost if permitted to run away, doubly lost if it carries also soil values and does destructive work below.

Experimental studies have shown that, on the average within our domain, *crops can use to advantage all the rainfall during the growing season,* and that, *in most cases, crops are the better for all the stored supplies that can be carried over from the non-growing seasons.* This greatly simplifies the general problem, for it justifies the conclusion—to which there are many local exceptions, of course—that the highest crop values will usually be secured when the soil is made to absorb as much of the rainfall and snowfall as practicable. In securing this maximum absorption and internal soil-work, the run-off, and hence the surface wash, will be reduced to a minimum. It has already been seen that the wash of even this inevitable minimum is likely to be still too great to keep the proper slow pace with soil-generation, when the surface has much slope . . . The practical problem then lies almost wholly in retaining and passing into the soil the maximum of the precipitation. Obviously, this gives the minimum of wash to foul the streams, to spread over the bottom lands, to choke the reservoirs, to waste the water-power, and to bar up the navigable rivers. *The solution of the problem for the tiller of the soil essentially solves the whole train of problems running from farm to river and from crop-production to navigation.*

LOOKING FORWARD

It must be stated that this circular does not undertake to tell the full story of the appalling wastage being caused by soil erosion. It merely refers briefly to some of the working processes of this greatest enemy to the most valuable asset of mankind (the agricultural lands), to some minimum estimates relating to the damage wrought, and to the meagerness of fundamental data concerning the problem. To visualize the full enormity of land impairment and devastation brought about by this ruthless agent is beyond the possibility of the mind. An era of land wreckage destined to weigh heavily upon the welfare of the next generation is at hand. Indeed, what has happened already and what is going on at an ever-increasing rate of progress is pressing upon many thousands of farmers now struggling to win subsistence from erosion-enfeebled soil. That the evil process is gaining momentum is due to the wearing away of the topsoil, which was more productive and more resistant to rainwash than the subsoil that is taking its place. That some 15,000,000 acres or more of formerly tilled land has been utterly destroyed by erosion in this country is but an insignificant part of the story, for it is the less violent form of erosional wastage, sheet erosion, that is doing the bulk of the damage to the land. Land depreciation by this slow process of planing off the surface is of almost incalculable extent and seriousness, and since the denudation does not cease when the subsoil is reached, there must be in the near future, unless methods of land usage are very radically changed, an enormous increase in the abandonment of farm lands.

A.—Properly constructed terrace in the Black Waxy belt of Texas. The sheet of water is flowing slowly and harmlessly off the field

B.—After its proper location has been determined, an efficient terrace can be plowed up without much difficulty

C.—Mountain land of the Blue Ridge destroyed after a few crops of corn. Such land should not be cleared

Destructive erosion on an unwisely cleared slope in the Blue Ridge part of the southern Appalachians

A.—Wind erosion, the result of drought and trampling, New Mexico. Excessive concentration of cattle near permanent water during drought periods results in serious trampling. Wind has removed 5 or 6 inches of soil, exposing the roots of the few remaining plants to desiccation
B.—Fire has destroyed the vegetation, and the steep slopes are rapidly washing away (California)

A.—A barren waste. Smelter fumes have destroyed the vegetation (Arizona)
B.—All the herbaceous vegetation has been destroyed by overgrazing. The brush and timber are insufficient to prevent serious sheet and shoe-string erosion and excessive soil deterioration

What would be the feeling of this Nation should a foreign nation suddenly enter the United States and destroy 90,000 acres of land, as erosion has been allowed to do in a single county? Any American of live imagination knows that the people of the United States would willingly spend $20,000,000,000 or as many billions as might be necessary, to redress the wrong. Because rain water was the evil-doer in this instance, which is but one of many, is the act forgiveable and is there no occasion for concern about it?

It is not necessary to go to China or to some other part of the world for examples of what eventually happens to unprotected slopes of cultivated areas. There is an abundance at home, not yet so vast in area as in China, but just as bad, and by no means small. It is well to observe, however, that millions of human beings have been driven out of the wasted uplands of China into the valleys of the great rivers, where the population is so dense and the land so completely used that even the roots of grain crops are dug for fuel. China cut the forests from the uplands and made no provision for protecting the bared slopes. Erosional débris sweeping out of these wasting highlands has rapidly extended the river deltas and made floods ever more difficult to control. After 4,000 years of building dikes and digging great systems of canals, the Yellow River broke over its banks and brought death to a million human beings during a single great flood. During one flood that great river, known in China as the "scourge of the sons of Han," changed its channel to enter the sea 400 miles from its former mouth.

No one, of course, wants anything remotely like this to take place in this country, but "coming events cast their shadows before." That the greatest flood of which we have reliable records came down the Mississippi in 1927 was a prophetic event. G. E. Martin's statement (*14*) about erosion as an enemy to agriculture—"It is very unlikely that any other industry could suffer such severe losses and survive"—is prophetic. That bare land at the Missouri Agricultural Experiment Station was found to be wasting 137 times faster than land covered with bluegrass, on a slope of less than 4 per cent gradient, is prophetic. That many millions of acres of cut-over land lie bare and desolate and exposed to the ravages of fire and erosion, with but pitifully little done toward reforestation, is prophetic. That minimum estimates show that the rate of plant-food wastage by erosion is twenty-one times faster than the rate at which it is being lost in crops removed, is prophetic.

These shadows are portentous of evil conditions that will be acutely felt by posterity. Shall we not proceed immediately to help the present generation of farmers and to conserve the heritage of posterity?

The writer, after 24 years spent in studying the soils of the United States, is of the opinion that soil erosion is the biggest problem confronting the farmers of the Nation over a tremendous part of its agricultural lands. It seems scarcely necessary to state the perfectly obvious fact that a very large part of this impoverishment and wastage has taken place since the clearing of the forests, the breaking of the prairie sod, and the overgrazing of pasture lands. A little is being done here and there to check the loss—an infinitesimal part of what should be done.

PART 2. SOIL EROSION ON WESTERN GRAZING LANDS

By W. R. Chapline

INTRODUCTION

The toll the West has paid to soil erosion is enormous. Countless slopes, once covered with rich soil and a dense carpet of herbaceous and browse plants capable of profitably supporting millions of cattle and sheep, have been so wasted by sheet and gully erosion following depletion of the vegetation that they can now support far less than half the number of livestock that once grazed upon them. Furthermore, the loss of the valuable surface soil and the exposure of the less productive subsoil have made difficult the reestablishment of an abundant stand of plants. Fertile valleys, their good soil cut away by silt-laden flood waters or covered with sand and gravel, have had their value seriously impaired or have even had to be abandoned. The silt has finally been carried to the mouths of rivers to clog channels and hamper navigation.

Water available for agriculture, power, and industry will largely determine the development and prosperity of the West. In 1920 approximately 19,000,000 acres in the West were under irrigation, and it was estimated that this area could be extended to 51,000,000 acres by the conservation and development of the entire water supply (*10*). Erosion, however, is already endangering established projects and making prospective ones uncertain. A more adequate protective covering is necessary to safeguard these watershed interests.

Erosion, the removal of the soil cover by water or wind, is taking place everywhere, but under favorable conditions nature starts at once rebuilding the soil from the decomposition of rocks or by the addition of humic material from decaying vegetable matter. Numerous factors, of which climate, soil, topography, and geologic formation are doubtless the most important, influence these processes, but the vegetative cover is the main single controllable factor. It is recognized that forest cover furnishes the greatest protective value for preventing erosion and soil washing and for regulating stream flow. It is, therefore, important that forest growth, including its understory of herbaceous and shrubby vegetation, be maintained wherever possible at the headwaters on all streams used for irrigation, power, or navigation. West of the one hundredth meridian, however, forests grow on only 13 per cent of the land area. It falls to herbaceous and shrubby vegetation to afford the necessary protection to the soil and stream flow on the remaining 87 per cent.

It is this vast area of unforested land that largely supports the range livestock industry which in turn plays a principal part in the prosperity of the West. On these western ranges the protective cover of vegetation varies from a very sparse stand of browse and annuals on desert areas to an almost complete cover of grass on untimbered alpine slopes where soil and moisture are favorable. The normal stand of native vegetation for the region as a whole is considered to be a stand covering 50 to 60 per cent of the soil surface.

EROSION AGENCIES ON RANGE LANDS

The natural balance on these arid and semiarid lands, between the forces that tear down and those that build up the soil, is a delicate one, but if the vegetative cover is not disturbed erosion is usually slight. Natural agencies may occasionally produce abnormal erosion. Man's activities, however, by reducing the vegetative cover or altering the topography or soil, can and do upset the balance completely. Where this occurs these activities are seldom abated, and severe erosion follows. Once started, this may develop to disastrous proportions.

NATURAL AGENCIES

Drought, the shortage of rainfall for a year or more, usually accompanied by excessive evaporation, may cause serious depletion of the vegetative stand. If so, it will probably be followed by wind or water erosion or both. (Pl. 11, A.) This is particularly true in the areas subject to a low average annual rainfall. On the Jornada Range Reserve in southern New Mexico, for example, drought caused a reduction of 40 per cent in the density of herbaceous vegetation on ungrazed areas between 1916 and 1919 (*11*) and a still greater reduction in 1922. A similar reduction of herbaceous vegetation was noted by Director Culley on the Santa Rita Range Reserve in southern Arizona, the loss being greater on ungrazed than on properly grazed range.

Cloud-bursts, snowslides, and excessive moisture on slopes, followed by landslides or landslips, may also unbalance the delicate equilibrium which nature has so carefully developed and may lead to excessive erosion. But nature, if unimpeded by man, quickly sets to work to reestablish the balance; the vegetation gradually spreads over the scars, the slopes assume an angle of repose, and eventually the normal conditions return.

FIRES

Of all the man-caused erosion, that resulting from fire has probably been the worst. Everywhere one goes throughout the West, old settlers tell of extensive fires allowed to burn over ranges and forests in the early days, little or no effort being made to stop them. Many of these fires were set for the purpose of exposing minerals or in the mistaken belief that they would improve the range. Even when great effort is made to control them, fires destroy the vegetation on hundreds of thousands of acres in the West every year. A single prairie fire observed in the sand hills of western Nebraska in the spring of 1910 burned the grass from an area 100 miles long and up to 30 miles in width. With recurring fires, the stand of vegetation becomes more open, the length of time the ground is bare increases, and erosion becomes more active.

In California, fires have completely or partially devastated more than 981,000 acres of forest, brush, and grass land. Of brush and grass land alone, they have burned over annually from 1916 to 1921 an average area of 336,000 acres. (*15*) In the southern California brushfields, one may see a 1,000-acre burn wherein practically all of the vegetation, which required years to grow, was destroyed in a few hours. When one stands in several inches of ashes on an extremely steep slope in the center of such a burn, knowing that the humus, "the life of the soil," has been consumed and that those ashes and the surface layer of soil will be washed down the slope with the first dashing rain, he realizes how great is the soil destruction from fire and erosion. (Pl. 11, B.)

MINING

Hydraulic mining along streams throughout the West tore out the valleys and sides of gulches, leaving the débris in the creek beds for high water to carry down to fill the lower river channels or to spread like a wasting hand over fertile farm land. This destruction has been especially severe in California but is now under regulation (*15*).

Smelters, an adjunct to mining, depleted several hundred thousand acres of western range land and completely destroyed the vegetation on a considerable part of it. (Pl. 12, A.) Bennett has already in this circular noted the excessive erosion on the denuded slopes. Damage is still being done, even though in some places recent abatement measures have reduced smelter-smoke injury materially.

OVERGRAZING

Following 1870 there was a rapid expansion of the range livestock industry in the West, especially in sheep raising. By the late seventies the expansion in cattle was in full swing and pressed on with prospects for a rich harvest (*1, 22, 23*). As numbers of livestock increased the palatable forage plants were grazed closer and closer, and their vigor was sapped. Instead of thick grass knee high, of which the early stockmen speak, there were shorter and sparser grass blades and stems; finally many of the plants gave up the struggle, and the stand was thinned. The less valuable plants were then grazed more severely, until they too were practically eliminated. The hungry animals in their search for feed trampled the range, destroying plant roots and packing the soil. The ranges became dust beds. Residents of Utah tell of being able to count the herds of sheep on the mountains by the dust clouds rising as the sheep trailed through the country. Under such conditions there was nothing to check the rain as it fell; the more compact soil could not absorb the water, which ran off and was quickly converted to a slimy mass of flowing mud. Shoe-string gullies started and speedily gained depth, while the main drainage channels became raging torrents. The rich friable surface soil was washed away and the heavy clay subsoil exposed.

Even before 1890 there were more livestock on the ranges than they could support satisfactorily. Oversupply brought a drop in values and caused a greater proportion of animals to be held on the range, until the drought of the early nineties, coupled with the greatly reduced feed supply, wiped out large numbers by starvation and forced stockmen to sell countless others at a loss.

Thus the stockmen have been, in part, victims of circumstance. Under "free," unregulated, and competitive grazing the range of any stockman who endeavored to have plenty of feed on his range would be eaten out by tramp stockmen who were eager to find such "pickings." Then, too, with drought reducing the feed supply 40 per cent or more in a year, it was impossible without control to make the adjustments necessary to assure reserve feed. When prices were high everyone desired to hold stock in the hope of profiting by the rise, and when prices broke there was at times no market whatever for the surplus animals. All of these factors tending to overstocking and overgrazing developed conditions unusually favorable for soil destruction, and there were few ranges of the West which did not suffer severely from erosion in the nineties, a considerable part of which is not yet checked.

The regulation of grazing within the national forests and the consolidation of private holdings have greatly reduced the area of range overgrazed, but extensive areas throughout the West are still deteriorating through excessive or otherwise improper grazing use. (Pl. 12, B.) The most serious situation at present is that on the 196,000,000 acres of unappropriated and unreserved Federal lands with their unfenced, intermingled State and private lands, on which grazing can not now be legally controlled. Much of this public domain lies in the foothills and should furnish the abundant spring and fall feed essential to profitable livestock production. Drought and overgrazing, however, are seriously impairing the feed and watershed values of large expanses of these important Federal lands by robbing them, through erosion, of the soil material necessary to maintain a protective covering.

OTHER AGENCIES

Other agencies, such as the building of roads and trails, the cultivation of steep slopes, the clearing of timber and brush from mountain-valley bottoms, the draining of wet meadows to facilitate cultivation, and the straightening or changing of stream channels, have come about with settlement and have caused erosion of range lands. These, with the exception of attempted hillside farming, have not as a rule caused sheet or shoe-string erosion of the slopes, but by cutting out the rich valley bottoms they have caused considerable damage. Rodents, mentioned earlier by Bennett, by denuding some range areas have thereby caused erosion.

The early roads and trails followed the watercourses through the foothills and mountains. The sod was broken on the slopes, so that water might accumulate and start cutting runways. (Pl. 13, A.) The natural beds of streams were altered to allow teams and wagons to pass, and the water was thus aided in its cutting process. Even to-day the construction of mountain roads without proper drainage is a common cause of unnecessary erosion. Bates (*2*) found, in logging the timber from a mountain wastershed in Colorado, that the scar left by the logging road was largely responsible for an increase of 519 per cent over normal in the sediment eroded from the watershed during the fourth year after cutting. Herbaceous and shrubby vegetation quickly reclaimed the rest of the area, however, and protected the slopes from undue erosion.

Long, narrow, fertile valleys occur throughout the mountain region. Farmers and ranchers have pressed farther and farther back into these valleys, cleared them of their timber and brush or turned under the sod, and cut ditches to drain the marshes, unknowingly removing the natural obstructions and reservoirs which slowed up the run-off. In floods these more effective drainage channels increased the carrying power of the water manyfold and the cutting power enormously. A slight lowering of the stream bed in the valley was enough to start the formation of deep, perpendicular-sided gullies or "arroyos," ever widening and deepening. (Pl. 13, B.)

Many of these arroyos are now 10 to 30 feet deep and 100 feet or more in width, while the larger streams have cut deeper and still wider drainage lines. Only an occasional rancher is doing anything to arrest this cutting and usually only enough to afford immediate protection to his farm land or improvements along the arroyo. As no real attempt is made to control the arroyo it continues extending its tentaclelike arms back and up the draws by accelerating the flow of water over the falls at each finger's end; and it saps the life from the valley and its tributaries by lowering the water table and tearing out the rich alluvial soil. Such a lowering of the water table has drained the valley meadows, forcing a replacement of valuable stands of grasses with a sparser stand of drought-resistant plants of lower value. Increasing flood waters rushing through the channels have cut away the adjacent agricultural lands, threatening their permanency and materially reducing their value. Ninety per cent of the farm land in the Blue River watershed of Arizona was destroyed by just such erosion, the damage over this relatively small area totalling $522,000.[1]

THE SERIOUS EFFECTS OF THE EROSION

The more rapid run-off from the depleted, untimbered slopes, the greater quantities of silt carried, the quicker concentration of high water from small drainages into flood crests that eventually assume disastrous proportions in the West, cause this depletion to play an important part in the distress occasioned in the main river valleys. The losses in the western range region are, however, more readily recognized and are in themselves of sufficient magnitude to demand urgent attention.

FLOOD DAMAGE TO MOUNTAIN VALLEYS

Olmstead (*17*) considered the ultimate destruction of the Safford Valley lands of Graham County, Ariz., by the Gila River floods so certain if not attended to, and so closely related to the serious erosion of slopes and valleys occurring in the upper watershed, that he recommended an expenditure by the Federal Government of $5,000,000, which should cover not only bank protection in the valley (pl. 14, A), but check dams and other erosion control and flood-prevention works on the watershed as well, even though the value of the land to be protected was at that time (1916) approximately

---

[1] UNITED STATES DEPARTMENT OF AGRICULTURE, FOREST SERVICE. WATERSHED HANDBOOK [FOR SOUTHWESTERN DISTRICT]. 28 p., illus. 1923. [Mimeographed.]

only $5,000,000. He emphasized the necessity of vigorously combating flood waters where they start and mastering them there by restoring the vegetative cover (pl. 14, B), by regulating grazing, and by putting in small check dams in the channels.

Of 32 agricultural valleys in the national forests of Arizona and New Mexico, only 3, or less than 10 per cent, had no erosion in 1923; 4, or 13 per cent, had been ruined by erosion, 31 per cent partly ruined, and of the remainder 28 per cent had noticeable erosion started. It is estimated that about 100,000 acres of agricultural land in these two States have been lost through erosion, nearly one-fourth as much as has been reclaimed by United States reclamation projects in the two States.[2]

In Utah, Reynolds (*20*) reported that previous to 1888 there was no record of serious floods from the Manti National Forest area. It was originally well protected by an abundant ground cover of shrubs and grasses, which delayed the run-off and prevented erosion. Heavy grazing by horses, cattle, and sheep, however, resulted in the almost complete destruction of this cover. Since 1888 floods of great violence and destructiveness have been common, and it is estimated that up to 1910 they caused a loss of approximately $225,000. This writer urges a restoration of the natural cover of shrubs and grasses and points out that the closing of the upper part of Manti Canyon to grazing in 1903 made it less subject to floods.

Hundreds of thousands of dollars are required yearly to repair and reconstruct roads, trails, bridges, and other public and private works destroyed by flood waters in valley lands in practically all parts of the West.

Furthermore much of the water which is so necessary for periodic irrigation and which should come gradually from underground sources is taken off in the flash floods, leaving irrigated farms to suffer from an inadequate water supply during the latter part of the irrigating period.

SILTING OF RESERVOIRS

One of the worst features of floods is the enormous quantity of silt carried from the slopes and rich valley bottoms along with rocks, bowlders, and other débris out on to valley farm lands (pl. 15, A) or into irrigation and other reservoirs. Talbot (*24*) reports an average deposit of 1 foot annually in 30 southwestern reservoirs used for livestock watering, making the average life for the reservoirs less than 15 years. Such a loss demands most urgent consideration of the silt problem. He recommends, in addition to check dams and other channel obstructions, the restoration of depleted vegetative covering on watersheds and the maintenance of as heavy a covering as possible, especially in the channels just above the reservoirs.

J. W. Taylor, manager of the Water Users' Association of the Elephant Butte project of New Mexico, states that 18,000 acre-feet of silt, equal to 0.7 per cent of its capacity, is entering the Elephant Butte Reservoir annually. While this may appear small, the 140 years which it might require to fill the reservoir with silt is but a short period in the life of the project. Within 50 years one-

[2] See footnote 1.

third of the capacity, the equivalent of a year's water requirements, will be occupied by silt, endangering the water supply for irrigation over a period of several dry years such as is common in the region.

For a number of years the rainfall on the watershed of the Roosevelt Reservoir has been below normal. In 1925 the water supply became very low. If the part of the reservoir capacity occupied by silt had been available for water, the shortage would not have been so great. During the dry period rains washed much of the surface soil from slopes into small gullies and occasionally as far as main channels. In 1927 torrential rains fell rapidly and, with the vegetation depleted by drought and overgrazing, caused terrific cutting of slopes. The several years' soil accumulations in the channels were carried farther down the main stream courses, usually to be dropped into the reservoir. Plate 15, B, shows strikingly the enormous quantity of silt, in the upper parts of the reservoir basin, being washed out by erosion of the silt banks and carried down nearer the dam. On experimental plots on the range where the vegetation had made normal growth the roots bound the soil, with the result that soil cutting was not so severe and much water-carried soil coming into the plots from above lodged in grass or brush.

REDUCED PRODUCTIVITY OF RANGE LANDS

In the midst of such great destruction by floods and silting to valley farm lands, irrigation and other reservoirs, roads, bridges, and other public works, the excessive damage to the productivity of the range lands for livestock grazing has hardly been appreciated. In their original condition the slopes and valleys, except in those arid parts where rainfall was very light, were well carpeted with valuable grasses and a small percentage of other herbaceous and shrubby plants. The decaying vegetable matter had built up the surface soil into a friable condition and added to it a large quantity of rich organic matter. The mulch of decaying vegetable matter acted as a sponge, and the friable humic character of the soil allowed a maximum moisture penetration. The result was that the forage plants made the most of the rainfall and the fertile soil and produced abundantly.

When erosion removes the top layer of soil it robs the plants. If erosion continues, the soil may become incapable of producing the stand it once supported. The experiments at the Great Basin Experiment Station (*21*) showed that noneroded soil was much richer than eroded soil in lime, phosphoric acid, and total nitrogen; that the water-holding capacity was greater; and that the water required by representative plants to produce a pound of dry matter was less. A great many more leaves, greater stem and leaf length, and more dry matter are produced on the noneroded than on the eroded soil, even with a notably smaller supply of water. The conclusion was drawn that erosion is detrimental to plant growth chiefly because it brings about two conditions of soil impoverishment: (1) Lack of adequate soil moisture for full plant development and seed production and (2) lack of adequate plant nutrients in the soil for good growth. Furthermore, reestablishment of the vegetative cover is made more difficult.

A.—Water accumulated in the wagon tracks and cut these gullies (Montana)
B.—Arroyo on Muddy Creek, Colo. The barn is about to be undermined

A.—Flood control fence constructed with iron posts to reduce velocity of flow, induce silting, and prevent bank cutting in a large drainage (Arizona)
B.—The establishment of grass in a gully tended to check further cutting (Arizona)

A.—Muck, driftwood, and other débris deposited in a young orange orchard by flood from hills made barren by fire
B.—Salt River, in slight flood stage, undercutting silt banks in the upper part of the Roosevelt Reservoir basin and carrying the mud down nearer the dam

Abundant excellent feed and a maximum of watershed protection (Montana)

Because of lowered moisture content from soil exposure and lowered water-holding capacity on eroded soil supporting a thin stand of vegetation, seed germinates poorly and many of the seedlings die.

In Montana W. G. McGinnies found that the blue bunch grass—sedge—weed type, the optimum herbaceous vegetation, normally covers 60 to 80 per cent of the soil surface and that but 2 acres of this cover are required to furnish a cow a month's feed. When the stand is thinned and the humus is washed out of the soil by erosion the valuable grasses give way and finally disappear, and plants of low value increase. Eventually, as the soil is depleted, rabbit brush—yellow brush—weed type, will take possession. This type usually covers but 20 to 40 per cent of the soil, and approximately 11 acres of it are required to furnish a cow a month's feed.

Numerous stockmen have remarked that the ranges of the West do not and will not carry as many animals as they once grazed. William Bailey, former tax commissioner of Utah, has stated that the "West Desert" of Utah will not support a tenth of what it once did. The reduced productivity of the ranges has increased the instability of the range livestock industry and made profitable production more difficult. Unless the number of livestock is reduced to conform to the smaller feed supply, the inevitable outcome is inadequate range feed, excessive death losses (particularly in drought periods), fewer calves and lambs, and decreased meat production. If the number is reduced, the per head investment in land and improvements may prove to be excessive.

WHAT NEEDS TO BE DONE

REESTABLISH THE VEGETATIVE COVER

This enormous wastage must be stopped. The problem is to determine and apply ways and means of checking the present extensive erosion, restoring the watersheds, preventing abnormal erosion, and obtaining permanent economic use of range land in the West. (Pl. 16.)

Of first importance is the effort to reestablish and conserve the optimum vegetative cover. Plants not only lessen the force of rainfall but intercept part of it. Vegetation improves soil structure, allowing greater moisture penetration; it increases the water-holding capacity of the soil by increasing organic matter; it breaks the effect of wind; it binds the soil and lessens sheet erosion; it obstructs run-off and reduces the velocity of flow and the carrying power of the water; and by catching soil particles it tends to form miniature terraces on slopes and dams and fills in small gullies. The more complete the plant cover, the more adequate is the protection against erosion.

If erosion is checked on a depleted area the vegetation present will gradually spread, slowly increasing the vegetable matter and plant foods in the soil. Short-lived species will be replaced by perennials and better soil-binding plants and, as the fertility of the soil is further improved, the more permanent type of perennial forage plants will become established.

The value of this in controlling erosion is indicated by the experiments on high mountain watersheds at the Great Basin Experiment

Station in Utah. After herbaceous vegetation had improved until it covered 40 per cent of the soil surface, the run-off from summer rains was 55 per cent less, and sediment eroded 56 per cent less than when the vegetation covered but 16 per cent of the surface. Run-off and erosion from melting snow appeared to be affected much less by the change in herbaceous vegetation. Though approximately 95 per cent of the annual run-off was from melting snow, it carried only 12 per cent of the sediment removed; the 5 per cent of run-off from summer rains carried 88 per cent of the sediment eroded annually. Thus, under the conditions prevailing in the experiment, the greatest need is for vegetation on the range to prevent the great erosion damage from summer rains.

Sometimes the range improvement is extremely slow, but every bit of progress helps not only in reducing erosion but in increasing the grazing value of the land. At the 1927 field day of the Great Basin Experiment Station it was shown that on an area on which the vegetation had been all but destroyed in 1903, the soil was so badly eroded and depleted that it had not yet built up to where it would support a stand of valuable forage species. Even so there had been a notable increase in carrying capacity. In 1927, 7.4 acres were required to support a cow for a month. On another area on which the stand of vegetation had been badly depleted but on which the soil was not so wasted, and adjoining a hillside where grass plants still remained under the protection of brush so that seed was available, a rather dense cover of valuable plants had become reestablished by 1927, and only 2.4 acres were required to support a cow a month. The greater part of this improvement has come in the last five years. The rate of improvement depends largely on the quantity of plants on the range which reseed or otherwise revegetate readily and on the methods of range management applied to aid recovery. In the Southwest rather badly depleted ranges can, with proper grazing, be reestablished, with three or four times as much forage as they are now supporting, in about five years.

The value of trees, shrubs, and grass for bank protection along small stream courses should be better appreciated, so that efforts will be made to protect them or to reestablish the stand if it has been destroyed.

REGULATE GRAZING

Stockmen are recognizing that conservative grazing keeps their animals in a good, thrifty condition throughout the year, increases the number of young produced, reduces death losses, and increases the weight and finish of salable animals, thus affording top prices and profitable production. Conservative grazing implies placing no more livestock on the range than the feed will support, and allowing the palatable plants to get enough of a start in the spring so that grazing will not impair their vigor. It also implies removing the livestock in the fall before the soil becomes so wet that trampling would injure it, unless the range and soil are of a character that allows yearlong use. A certain reserve of feed as an insurance against the ever-recurring drought is essential to assure sustained livestock production and watershed protection. Also, distribution of livestock over the range so as to obtain as even a use as practicable without undue concentration or trampling is important.

There are areas on which erosion injury is now occurring that could be greatly improved if grazed by a different class of livestock. This is true of certain rough mountain ranges where attempts to get full utilization of the feed on the range as a whole by cattle grazing is causing undue concentration in valleys and consequent erosion damage. Sheep would use the slopes to better advantage and could be more easily held off the damaged valley areas. Certain brush areas could be grazed more profitably by Angora goats than by the cattle and sheep now grazing there, bringing about improvement in the watershed and erosion conditions. Likewise, on some areas a change from sheep to cattle would prove desirable.

Deferred grazing or deferred and rotation grazing, which provides for reserving grazing from part or all of areas of range land until after seed maturity, although now widely applied, deserves more extensive use throughout the West. Other similar improved systems adapted to specific soil and forage types need to be devised or further developed so as to improve range lands without loss of use of the range forage. Experiments at the Great Basin station and on the Jornada and Santa Rita Range Reserves in the Southwest indicate that, except where the vegetative stand has been practically eliminated and the exposed soil is seriously eroded, the native cover can ordinarily be restored under properly regulated livestock grazing almost as well and as quickly as under total protection from grazing. Furthermore, the range can more easily be maintained at its best with grazing than without.

Some form of control of the unappropriated public domain is essential if it is to be restored. Most stockmen will agree that it should probably take the form of Federal regulation.

PROTECT THE COVER AGAINST FIRE

In recent years there has been a great awakening of the public to the need for fire prevention and quick suppression, especially in timber and brush lands. Most Western States have stringent laws regarding carelessness with fire, but, as brought out above, 336,000 acres of brush and grassland are burned yearly in California alone. The possible erosion danger should be carefully considered before fire is set with a view to improving the range. Under regulation, excessive grazing is sometimes practiced as a fire-prevention measure; but this in turn causes erosion, reduced feed values, and usually uneconomic livestock production. With depleted vegetative stands the fire danger is not great, but as the vegetation on range lands is improved for watershed protection it will be necessary to give more attention to fire protection.

AID EROSION CONTROL BY ARTIFICIAL MEANS

Artificial reseeding of range lands to known cultivated forage plants has not proved practicable except on areas with unusually favorable soil and moisture conditions. Tests are being made, however, of a number of species that show some promise. These tests deserve considerable expansion. In the meantime, management of the native vegetation so that it may serve to best advantage will be the main means of revegetating depleted areas. Of course, on important watersheds where the value of the stand is not alone determined by its grazing value it may pay to seed for erosion control.

Sampson and Weyl (*21*) found terracing and planting of steep, barren, eroded, high-mountain hillsides to native plants possible but costly. They urge such methods on areas that have eroded to such a point that natural revegetation is extremely slow and the vegetation present noneffective in binding the soil and in preventing erratic run-off.

Engineering works for the control of erosion have been used extensively in the mountain regions of Europe. Large dams for holding back flood crests or even for the catching of excessive silt have been established in a few places in the West. Check dams, small structures of rocks, logs, brush, or other materials, have also been constructed in the smaller stream channels, largely for the purpose of reducing the velocity of the water and thereby its carrying and cutting power. Such works are usually costly and unless they are supplemented with the maintenance of the highest type of vegetation the land is capable of supporting are apt to prove ineffective. As the importance of erosion control is recognized, however, engineering works, especially check dams, will doubtless come into greater use.

CONCLUSION

With erosion losses on western grazing lands so great, corrective action must be taken soon if far greater damage and more difficult control are to be obviated. Owners of range land should consider the use of their land not alone for immediate gain, but still more in the light of the future productivity of the range, the protection of water supply, and stream-flow regulation. Overgrazing should be stopped at once; control or regulation of the badly abused unappropriated and unreserved public domain should not be longer delayed by the Federal Government. Arroyo cutting must be checked by engineering works and the establishment of vegetation in the bed and on the sides of the arroyos. Range landowners, irrigationists, and the State and Federal Governments should band together to use every available means for checking erosion, floods, and inadequate water supply at their source, on the slopes, in gullies, and on small drainages of the watersheds. The Federal Government has a direct responsibility since Federal lands occupy such a large part of the West.

The main obstacle to action and one that has greatly delayed remedial measures has been lack of information as to the seriousness of the situation and as to concrete things which should be done under specific conditions. Without this information it is possible to work only in a broad way, rather than to attain a permanent control of erosion on range lands in a really constructive and economical manner. In view of the important part that herbaceous and shrubby vegetation play in controlling erosion of such lands it is essential that research determine just what is the optimum stand of vegetation that can be made to grow on the widely varying soil types and under the extreme climatic conditions of the West and the influence of this vegetation on water supply. It is equally important to know more concretely just what grazing use can be allowed under each of the main range and watershed conditions to assure profitable livestock production and a maximum of protection to the soil. Along these lines research is already doing its best to help the stockman and range landowner, but hardly more than a start has as yet been made. Faced with so big a problem, research needs the most earnest encouragement and support.

LITERATURE CITED

(1) Barnes, W. C.
1925. THE STORY OF THE RANGE. U. S. Congress, 69th, 1st sess., Hearings Senate Res. 347, pt. 6, p. 1579–1640, illus.

(2) Bates, C. G.
1924. THE EROSION PROBLEM. Jour. Forestry 22: 498–505.

(3) Bennett, H. H.
1921. THE SOILS AND AGRICULTURE OF THE SOUTHERN STATES. 399 p., illus. New York, The Macmillan Co.

(4) Carr, M. E., Welsh, F. S., Crabb, G. A., Allen, R. T., and Byers, W. C.
1914. SOIL SURVEY OF FAIRFIELD COUNTY, SOUTH CAROLINA. U. S. Dept. Agr., Bur. Soils, Field Oper. 1912, Rpt. 13: 479–511, illus.

(5) Chamberlin, T. C.
1909. SOIL WASTAGE. *In* Proceedings of a Conference of Governors in the White House, Washington, D. C., 1908. U. S. Congress, 60th, 2d sess., House Doc. 1425, p. 75–83.

(6) Dole, R. B., and Stabler, H.
1909. DENUDATION. U. S. Geol. Survey Water-Supply Paper 234: 78–93.

(7) Duley, F. L.
1924. CONTROLLING SURFACE EROSION OF FARM LANDS. Missouri Agr. Expt. Sta. Bul. 211, 23 p., illus.

(8) Geib, H. V.
1926. SOIL SURVEY OF ROCKWALL COUNTY, TEXAS. U. S. Dept. Agr., Bur. Soils, Field Oper. 1923, p. 123–152, illus. (Advance sheets.)

(9) Gilbert, G. K.
1914. THE TRANSPORTATION OF DÉBRIS BY RUNNING WATER. U. S. Geol. Survey Prof. Paper 86, 263 p., illus.

(10) Gray, L. C., Baker, O. E., Marschner, F. J., Weitz, B. O., Chapline, W. R., Shepard, W., and Zon, R.
1924. THE UTILIZATION OF OUR LANDS FOR CROPS, PASTURE AND FORESTS. U. S. Dept. Agr. Yearbook 1923: 415–506, illus.

(11) Jardine, J. T., and Forsling, C. L.
1922. RANGE AND CATTLE MANAGEMENT DURING DROUGHT. U. S. Dept. Agr. Bul. 1031, 84 p., illus.

(12) Long, D. D., Beck, M. W., Hall, E. C., and Burdette, W. W.
1916. SOIL SURVEY OF STEWART COUNTY, GEORGIA. U. S. Dept. Agr., Bur. Soils, Field Oper. 1913, Rpt. 15: 545–606, illus.

(13) Lowe, E. N.
1922. REFORESTATION, SOIL EROSION AND FLOOD CONTROL IN THE YAZOO DRAINAGE BASIN. South. Forestry Cong. Proc. 4: 10–11.

(14) Martin, G. E.
1927. TERRACING IN OKLAHOMA. Okla. Agr. Col. Ext. Co. Agt. Work Circ. 218, 19 p., illus. (Revised ed.)

(15) Munns, E. N.
1923. EROSION AND FLOOD PROBLEMS IN CALIFORNIA. 165 p., illus. Sacramento. (Calif. State Bd. Forestry Rpt. to Legislature 1921 on Senate Concurrent Res. 27.)

(16) National Industrial Conference Board.
1926. THE AGRICULTURAL PROBLEM IN THE UNITED STATES. 157 p., illus. New York.

(17) Olmstead, F. H.
1919. GILA RIVER FLOOD CONTROL. U. S. Congress, 65th, 3d sess., Senate Doc. 436, 94 p., illus.

(18) RAMSER, C. E.
1917. PREVENTION OF THE EROSION OF FARM LANDS BY TERRACING. U. S. Dept. Agr. Bul. 512, 40 p., illus.

(19) ———
1924. TERRACING FARM LANDS. U. S. Dept. Agr. Farmers' Bul. 1386, 22 p., illus.

(20) REYNOLDS, R. V. R.
1911. GRAZING AND FLOODS: A STUDY OF CONDITIONS IN THE MANTI NATIONAL FOREST, UTAH. U. S. Dept. Agr., Forest Serv. Bul. 91, 16 p., illus.

(21) SAMPSON, A. W., and WEYL, L. H.
1918. RANGE PRESERVATION AND ITS RELATION TO EROSION CONTROL ON WESTERN GRAZING LANDS. U. S. Dept. Agr. Bul. 675, 35 p., illus.

(22) SHEETS, E. W., BAKER, O. E., GIBBONS, C. E., STINE, O. C., and WILCOX, R. H.
1922. OUR BEEF SUPPLY. U. S. Dept. Agr. Yearbook 1921: 227–322, illus.

(23) SPENCER, D. A., HALL, M. C., MARSH, C. D., COTTON, J. S., GIBBONS, C. E., STINE, O. C., BAKER, O. E., VALGREN, V. N., JENNINGS, R. D., HOLMES, G. K., BELL, W. B., and BARNES, W. C.
1924. THE SHEEP INDUSTRY. U. S. Dept. Agr. Yearbook 1923: 229–310, illus.

(24) TALBOT, M. W.
1926. RANGE WATERING PLACES IN THE SOUTHWEST. U. S. Dept. Agr. Bul. 1358, 44 p., illus.

ADDITIONAL COPIES
OF THIS PUBLICATION MAY BE PROCURED FROM
THE SUPERINTENDENT OF DOCUMENTS
U.S. GOVERNMENT PRINTING OFFICE
WASHINGTON, D. C.
AT
25 CENTS PER COPY
▽

Spotted something that isn't quite right?

Let us know and we'll fix it for future reprints.
And as a thank you for your diligence, we'll
send you a free Sothis book of your choice.

publisher@sothis.press

This page intentionally left blank

Except of course for that line above. And this one.
And whilst we have your attention, perhaps you'd
like to follow us on Instagram @sothis.press. Go on.